PRINTED CIRCUIT BOARD BASICS

SECOND EDITION
BY MICHAEL FLATT

An Introduction to the PCB Industry

From the Publishers of
Printed Circuit Fabrication Magazine

mf Miller Freeman
A MEMBER OF THE UNITED NEWSPAPERS GROUP
San Francisco & Atlanta

Miller Freeman Inc., 600 Harrison Street, San Francisco, CA 94107
Publishers of *Printed Circuit Fabrication, Printed Circuit Design,* and
Circuits Assembly magazines.

Developmental Editor: Donna Esposito
Copy Editor: Susan Bureau
Technical Artist: Rick Eberly
Designer: Matt Kelsey
Photography by Wil Jones of Jones and Co., Phoenix, Arizona

ISBN 0-87930-232-1
Library of Congress Catalog Card Number: 91-61848
Printed in the United States of America
       93   94   95   96   5   4   3   2

Contents

Acknowledgments

The author wishes to thank those who contributed to the gathering of information for this book. Of particular note are Harvey Miller of Kirk-Miller Associates, David Bergman and Ray Pritchard of the IPC, Larry Velie of Velie Circuits, Bob Lutz of Continental Circuits, Bob Gonzalez of Quality Printed Circuits, and Kirk Lockett of Du Pont Electronics. Many thanks to all.

Also of note are the folks at Continental Circuits Corp.—Jack Bramel, Tony Gueli, Tom Linnen, John Maddux, Bill Moore, Norm Pettett, Lee Small, Tom Steele, Denise Turberville, Bob Watkins, John Whitley, and Rod Winn. Thanks for answering the many requests for facts and figures. Thanks also to those at Continental who gave such terrific assistance during the shoot for the photos contained herein.

Special thanks also to Donna Esposito and Susan Bureau of *PC FAB* magazine. Under their guidance and tutelage, rewrites, changes, and modifications were almost fun.

MOF

Introduction

To begin, set aside the fact that a printed circuit board (PCB) isn't really printed, isn't a circuit, and, actually, isn't a board either. A PCB is a custom device, designed by one company for a specific product, yet often manufactured by another. Add to this that the PCB is frequently assembled by a third company. Finally, it's the last part of an electronic product to be designed, but is the first part needed to start assembly!

Also consider that the manufacturing sequence for a PCB consists of more than 100 individual steps, involves a variety of raw materials, and employs mechanical, chemical, electrical, and photographic processes. So many different finishes, coatings, sizes, and thicknesses are available that they defy quantification and complicate uniformity of description.

It's not surprising that a person who wants to find basic, general information on printed circuit boards is hard pressed to do so.

When one discusses basic, general circuit board information, the task is to be complete without being complex, to explain and inform, but not to engulf the reader with a tidal wave of information.

This book has been structured to provide an introductory block of information to the student, the new employee, the circuit board buyer, the inspector, or anyone who needs to learn about PCB fabrication in a simple, quick, and clear manner. This book does not attempt to address all the material, equipment, and process variances. Rather, its focus is on the most widely accepted and utilized manufacturing methods.

To those whose preferred method, process, or equipment has been overlooked, the author offers a basic, general apology.

Michael O. Flatt
Mesa, Arizona

CHAPTER 1
Industry Overview

Printed circuit board. Printed wiring board. Bare board. Circuit board. Circuit card. Printed board. PCB. PWB. All of these names are used to describe a device that provides electrical interconnections and a surface for mounting electronic components. Although "printed wiring board" is more technically correct, the term "printed circuit board" (PCB) is most commonly used. The **Terms and Definitions** section of this book supplies concise differentiations, but all terms are used interchangeably.

EVOLUTION

PCBs came into use after World War II. Dr. Paul Eisler, an Austrian scientist working in England, is usually credited with making the first PCB. The concept was to replace radio tube wiring with something less bulky.

In the 1950s and '60s, boards with circuitry on one side (single sided) were the dominant variety. Still in use today, single-sided boards are the simplest variety of PCB. They are manufactured in high volume, most often for consumer electronics, and are the least expensive to produce.

During the late '60s and early '70s, processes were developed for plating copper on the walls of the drilled holes in circuit boards, allowing top and bottom circuitry to be electrically interconnected. Double-sided boards quickly became the industry standard. Used in more sophisticated consumer prod-

Bare (unassembled) PCB (rear) with assembled PCB in foreground

ucts and extensively in computer peripheral equipment, double-sided boards are more expensive than single-sided boards.

As the densities and complexities of electronic components increased, the multilayer board—a process of sandwiching several circuitry layers together—was developed. By the mid-1980s, multilayers accounted for the majority of U.S. output. Today's computers, aerospace equipment, and instrumentation and telecommunications gear all contain multilayer boards. Multilayers are the most expensive type of PCB to produce.

THE INDUSTRY

The U.S. PCB industry experienced monumental growth from the mid-1970s through the '80s. In 1975, the U.S. output was approximately $1 billion. This figure grew to $2.6 billion in 1980, $4.1 billion in 1985, and almost $5.5 billion in 1990. The worldwide output for PCBs in 1990 was $20.2 billion.

In the mid-1980s, there were approximately 2,000 PCB producers in the U.S. By 1990, that figure had dropped to approximately 900, due to high capital equipment costs, environmental regulation expenses, and offshore competition.

Most shops in the U.S. are privately owned and are called "independents" or "merchants." These shops manufacture PCBs to order for original equipment manufacturers (OEMs) or other users of PCBs. Some OEMs manufacture PCBs for their own internal consumption. These facilities are called "captive" shops. In 1979, captive producers accounted for 60% of the output, but since then, captive production has steadily declined. In 1990, more than 65% of the total U.S. output was manufactured by merchant producers.

PCB facilities can be small, with sales of a few thousand dol-

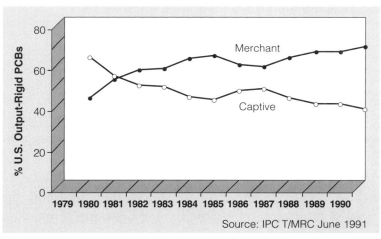

Figure 1-1. Merchant vs. Captive Output

Table 1-1. Top Ten Merchant Producers in North America, 1990*
(Output of printed circuit boards given in millions of dollars per year)

Company	Output
Hadco	158
Photocircuits	125
Diceon Electronics	113
Zycon	108
Circo Craft	84
Advance Circuits	83
Tyco	66
Tektronix	61
Sanmina	61
Continental Circuits (AZ)	60

Table 1-2. Top Ten Captive Producers in North America, 1990*
(Output of printed circuit boards given in millions of dollars per year)

Company	Output
IBM	418
AT&T	195
GM Hughes/Delco	153
DEC	125
Hewlett-Packard	68
Unisys	55
Texas Instruments	50
Raytheon	35
Rockwell	24
Thompson	24

*Source: Kirk-Miller Associates, Palo Alto, CA, 1991.

lars each month, or very large, with monthly sales in the millions. Table 1-1 shows the top 10 merchant producers; Table 1-2 shows the captive producers.

THE MARKET

The largest consumer of printed circuit boards is the computer industry. This dominance has been consistent for more than 10 years, but telecommunications and the automotive market have recently made significant increases.

Independent PCB manufacturers generally do not design or specify the boards they produce. Information regarding the shape of the board, the mechanical and electrical properties, the surface finish, and the material composition is provided by

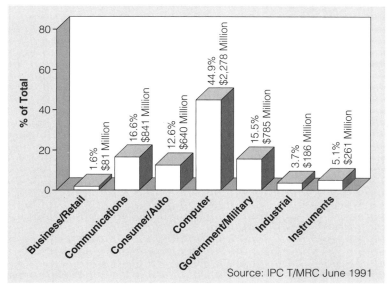

Figure 1-2. U.S. Market for PCBs, 1990

the PCB designer. The designer works either for an OEM, a design service bureau, or a contract manufacturer.

Before 1980, OEMs specified and ordered the vast majority of PCBs. Since then, however, the trend has been for OEMs to order the entire circuit board assembly from a contract manufacturer (or contract assembler). In this case, the specification of the bare board and the selection of the board fabricator is often left to the subcontractor, not the OEM. In 1990, contract assemblers purchased 8.3% ($453 million) of the boards produced in the U.S. Some industry experts predict that by the year 2000, this figure could be as high as 50%, making the contract assembler a major force in the electronics industry.

The expectations of the PCB consumer have changed dramatically. Customers are demanding higher quality products at more competitive prices. Forward-thinking PCB facilities are implementing philosophies and regimes such as total quality

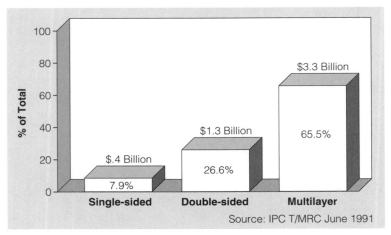

Figure 1-3. U.S. Output of Rigid PCBs, 1990

management, just-in-time inventory control, statistical process control, and design of experiments.

GENERAL DESCRIPTION

Most printed circuit boards are made from laminate, a flat, rigid material constructed with epoxy glass in the middle and copper foil on the outside. The epoxy glass serves as an insulating material and provides the structural strength for mounting components. The copper is the conductive medium through which electrical currents travel. The fabrication process begins with the laminate (also called the dielectric or the substrate) as the primary raw material.

Electrical interconnections are accomplished with copper traces (conductors, runs, circuitry) on the base material. The traces are made by selectively removing portions of the copper foil. Electrical current is also carried by copper deposited on the walls of the holes drilled in the boards, thus connecting the top surface circuitry to the bottom as well as to layers of circuitry inside the base material.

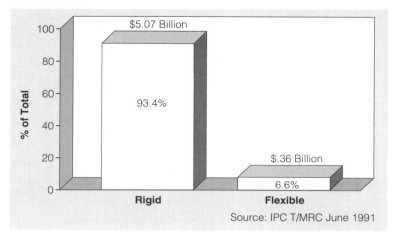

Figure 1-4. U.S. Output, Rigid vs. Flexible PCBs, 1990

Boards with circuits on one side are called single-sided boards; those with circuitry on both sides are called double sided. Boards with circuitry on both sides plus layers of circuit inside the base material are called multilayers. Multilayers are further distinguished by the number of layers.

The majority of boards produced in the U.S. today are $\frac{1}{16}$" thick. Some are as thick as $\frac{1}{4}$" or as thin as 0.005". A very thin PCB with the capability of bending is called a flexible circuit. However, more than 93% of U.S. production in 1990 was in rigid PCBs.

Components are mounted to the PCB in two ways:

1. Component legs are inserted through holes in the PCB and attached to pad areas on the opposite side of the board with solder. This method is called pin-in-hole or through-hole technology.

2. Component legs are soldered directly to pad areas on the surface of the board. This technique is called surface-mount technology (SMT). Surface-mount technology has

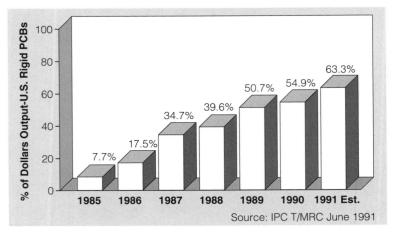

Source: IPC T/MRC June 1991

Figure 1-5. Growth of Surface-Mount Technology

been growing rapidly since 1984 as the assembly technique of choice in the U.S.

SUMMARY

Regardless of the name used, printed circuit boards provide a cost-effective, reliable method for interconnecting electronic components. PCBs can be small enough to fit in a hearing aid or large enough to drive a powerful computer. They can be rigid, flexible, or even three-dimensional.

The pressures on U.S. board makers are numerous. In addition to foreign competition, high capital equipment costs and increased environmental compliance standards are making the PCB business not only very price competitive, but also extremely difficult to enter.

There is no danger of PCBs being replaced in the next 10 years, however. More designs will call for semiconductor chips mounted directly to the board surface (chip on board). Multichip modules—several chips interconnected on a single substrate—are also emerging as a viable packaging technology, but their replacement impact on PCBs will be minimal.

CHAPTER 2
PCB Specifications

A specification is a document that defines and quantifies a PCB's mechanical, chemical, and electrical properties. Over the years, specifications have been generated that define raw materials, appearance, workmanship standards, and methods of testing. There is even a specification that details the data format for electronically describing a PCB.

Specifications serve to set standards and to establish uniform quality for the finished product. Private, public, and governmental agencies have expended thousands of hours of effort to create the various specifying documents. Four organizations are recognized as the primary sources of initiation, implementation, and control of PCB specs.

The Institute for Interconnecting and Packaging Electronic Circuits (IPC)

This international trade association of PCB designers, fabricators, and assemblers was founded as the Institute of Printed Circuits in 1957. In the years since, the IPC has taken the leadership position in generating new specifications for raw materials, processing, testing, and overall acceptability of PCBs.

Department of Defense, U.S. Government

For many years, the definitive documents for specifying PCBs were issued and controlled by various organizations within the military, hence the general name "MIL-SPECs." Today, the

Defense Electronics Supply Center (DESC) is the responsible government agency.

The National Electrical Manufacturers Association (NEMA)

This group was the first to issue quantified specs for laminates. The terms "FR4" and "G10" are NEMA grade designations.

Underwriters Laboratories (UL)

An independent testing laboratory, UL performs site inspections and laboratory testing to verify that fabricators' processes do not degrade the base laminate or impact the board in any way that would render the PCB unsafe.

A recognized UL symbol, similar to a company logo, indicates that the PCB made by a particular fabricator from a particular laminate subjected to a particular sequence of processes meets certain safety test criteria. Also, UL tests and issues flammability ratings for each laminate/process combination. These four ratings (94HB, 94V-2, 94V-1, and 94V-0) indicate how well a particular combination supports combustion (burns). The 94V-0 rating means virtually no support of combustion, while, at the opposite end of the spectrum, 94HB indicates non-flame-retardant properties.

Since most fabricators offer a variety of laminates and multiple processes, a particular board shop will have several UL-approved recognition symbols. UL publishes a listing of all UL-approved fabricators plus a further listing of each fabricator's recognized symbols and the related test data summary. This publication is the "Underwriters Laboratories Component Directory."

In addition, most fabricators have UL-issued cards that list their particular symbols and test ratings. These yellow cards are often the first document requested from a fabricator by a potential customer. Most customers insist that the fabricator's recognized UL symbol appear on the finished board.

Through the cooperative efforts of the agencies involved, the specifications for PCBs are now generally accepted and understood by board designers, fabricators, and users.

LAMINATE SPECS

Laminate is the primary raw material in PCB fabrication. Specification of laminate is accomplished in Military Standard MIL-P-13949G ("Military specification, plastic sheet, laminated metal clad, November 1988"). The IPC has also issued specifications for laminate (IPC-L-109B and IPC-L-1158).

Although the IPC is assuming more responsibility in the issuance of PCB specs, the military specification for laminate is still most widely used. The 13949 spec has served for many years as the predominant specification for commercial PCBs as well as military boards.

The laminate specifications define the electrical, mechanical, and chemical properties of the raw material. They cover all varieties of laminate from the most predominant, copper-clad epoxy glass, to those with unique electrical properties, such as polyimide and teflon.

The copper foil laminated to the epoxy glass is available in different thicknesses. The foil thicknesses are expressed in ounces meaning that a square foot of a given thickness of foil would weigh that many ounces. For instance, 1 sq. ft. of copper foil designated "1-oz. foil" would weigh 1 oz. The actual thickness of 1-oz. foil is 0.0014". Foils are specified as fractions (½ oz., ¼ oz., ¾ oz.) or whole numbers (1 oz., 2 oz., 3 oz.).

TYPICAL DESCRIPTIONS

A typical description of a copper-clad laminate is 0.062 1/1 FR4. This laminate is 0.062" thick, is clad with 1 oz. of copper on each side, and is NEMA grade FR4 epoxy glass. The example 0.031 2/0 FR4 has an overall thickness of 0.031", 2 oz. of copper on one side and no copper on the other, and is NEMA grade FR4.

The 13949 spec details additional physical characteristics. Each portion of the designation quantifies some parameter of the raw material. For example, the call out "GFN062 D1/D1 A1A" would quantify:

GFN	0.062	D1/D1	A	1	A
Base class material	Nominal base thickness	Type and nominal weight of foil	Grade of pits and dents	Class of thickness tolerances	Class of bow and twist.

OTHER SPECS

Specifications also exist that set forth guidelines for the PCB designer. The current document is MIL-STD-275E, amended July 1986, "Military Standard for printed wiring for electronic equipment." However, as is the case with several military specifications, this document is about to be superseded by an IPC-issued document. The new design document, IPC-D-275 "Design standard for rigid printed boards and rigid printed board assemblies," is due to be issued in late 1991.

Workmanship standards, which quantify finished board attributes, are found in MIL-P-55110D, amendment 4, issued in December 1984 as "Military specification, printed wiring boards, general specification for." Again, an IPC document, IPC-RB-276 "Qualification and performance specification for rigid printed boards," is near release. In addition, IPC-A-600D "Acceptability of printed boards" gives visual acceptance guidelines that are widely used by fabricators and users of PCBs.

The printed circuit board user (the OEM or assembler) often creates in-house documentation or internal specifications that quantify its own particular needs. These in-house specs may reference IPC documents, military standards, and UL ratings in addition to the particular workmanship, acceptability, and other criteria that the company feels are critical.

For additional information, review the following documents:

Military

The Defense Electronics Supply Center (DESC), Dayton, OH 45444-5288, 513/296-6278.

> MIL-P-13949—material
> MIL-STD-275—design criteria
> MIL-P-55110—workmanship/performance.

IPC

7380 North Lincoln Avenue, Lincolnwood, IL 60646, 708/677-2850, fax 708/677-9570.

> IPC-L-109B—material
> IPC-L-1158—material
> IPC-D-275—design criteria
> IPC-RB-276—workmanship/performance
> IPC-A-600—acceptability.

UL

The Underwriters Laboratories, P.O. Box 14995, Research Triangle Park, NC 27709-3995, 919/549-1400.

> UL-94—flammability
> UL 796—PCBs in general.

CHAPTER 3

Designing and Defining PCBs

The PCB design starts when the designer selects the electronic components to perform the required functions. Then, the task is basically to determine the most efficient and effective way to electrically interconnect the devices.

Before the 1980s, PCBs were designed or "layed out" manually by placing black tape and templates on clear mylar film. The original was two or four times larger than the PCB's actual size. This taped artwork was then photographically reduced, and the result was a film copy of the PCB pattern at actual size (artwork master).

Today's PCB designs are created on computers using computer-aided design (CAD) software programs specifically for PCB layout. CAD software may be simple enough to operate on a personal computer or may require the speed and storage capacity of an expensive workstation.

Electronic data describes the locations of holes and conductors in terms of an X and Y coordinate system. This data (called CAD data) can be used to generate photographic images of the PCB and to produce other manufacturing tooling items. Thus, the CAD data can be used in many areas of the factory, saving time and improving accuracy and quality.

Whether simple or complex, manual or CAD generated, the PCB design will provide the board fabricator with a great deal of

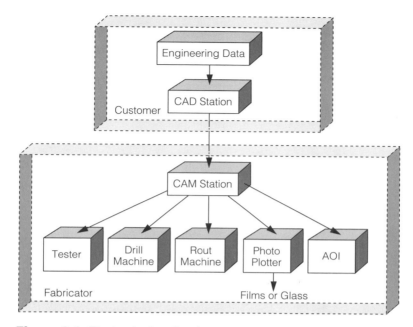

Figure 3-1. Electronic data flow in computer-integrated manufacturing (CIM)

manufacturing information, including the board dimensions, hole sizes, hole locations, and overall mechanical definition. Also incorporated are notes that reference

- The type of material
- Workmanship standards, internal or industry specifications
- Plating finish
- UL requirements
- Solder mask requirements
- Test requirements
- Special requirements.

The board layout data and manufacturing information are stored on magnetic tapes or transmitted by modem from the designer to the board fabricator.

Computer-aided manufacturing (CAM) workstation

The board fabricator receives the electronic data and can manipulate it with computer-aided manufacturing (CAM) workstations. Specially designed software programs allow the fabricator to perform panelization, generate drill and rout tapes, generate data to be used in electrical test and optical inspection, and transfer the design data to laser photoplotters to generate photographic images of the PCB on film or glass.

Until the early 1990s, manufacturers and engineers of CAD/CAM systems, photoplotters, and optical inspection and test equipment designed the data flow and format to suit their own needs. Thus, different kinds of equipment used different kinds of data formats. Now the IPC has developed specifications to fulfill the need for a universal data format. The main

Documentation package: punched paper tape, mechanical drawing, and mylar films of outerlayer circuitry. Not shown are innerlayer films, solder mask films, and component legend films.

document is IPC-D-350C "Printed board description and digital form."

DOCUMENTATION AND TOOLING

The manufacturing process begins with the manipulation of the data, hence the name "front end" given to CAM systems. The front-end systems convert customer information into manufacturing tooling. Tooling is the general name given to those items unique to a PCB design that are necessary for manufacturing, such as films, drilling and routing data, and test fixtures.

Glass phototooling

Phototooling

The first step in the tooling cycle is panelization. Multiple images of the same PCB are generated to economically fill the laminate panel. The computer "steps and repeats" the PCB image to the desired configuration.

Auxiliary features—layer numbers, UL symbols, borders, test coupons—are added to the panel. Panelization for each layer of circuitry plus those required for solder mask and component legend are prepared in the same way. The panelization is electronically stored until needed.

The panelization data is transferred to the laser photoplotter, a machine that "draws" the panelized image using laser light. The image is drawn on photographic film or sensitized glass plates.

The films, or glass, are the master phototooling (or production master). Some fabricators use the master directly in production to image the boards. Other companies use the master to produce more copies (working film) that are then used for production phototooling.

The use of glass plates in PCB production increased in the late 1980s with the increase in SMT designs. Glass is more dimensionally stable than film, which shrinks or grows with changes in temperature and humidity. On the other hand, glass is more expensive and harder to handle and store.

Drill and Rout Data

The next step in the tooling cycle is to generate the drill and rout data, either on punched paper tape, floppy disk, magnetic tape, or in computer memory.

Drill data is the information that tells the computer-controlled drilling machine the location and size of the holes in the PCB. Included in the data are the machine commands to change drill bits, vary spindle feed (rpm), and control the up and down rates (upfeed and downfeed) of the spindles.

Punched paper tapes are commonly used to get the information to the drilling area. But installing a computer network allows the data to be downloaded to the machines (or special data handlers), thus eliminating handling and storage of drill tapes.

Rout data, generated in the same way, is the information that instructs the computer-controlled router in machining the finished board dimensions. Again, direct downloading from the CAM workstation to the routers or to computer networks is gaining popularity.

PCBs partially routed and scored. Panels are shipped to customers in this form for assembly efficiency.

Inspection and Test Data

Before the 1980s, inspection of PCBs for flaws was performed visually under magnification. As the width of the conductors and the spaces between them decreased, it became necessary to use special measures to find breaks or shorts in the finer lines. This technology is called automatic optical inspection (AOI).

AOI gained almost immediate acceptance during the 1980s. Board fabricators would program the AOI memory using a known good board (golden board) as a reference. The machine would then compare what was in memory to what it was "seeing" as it scanned the board being inspected. In the case of a

Automatic optical inspection (AOI)

mismatch, the machine would identify the location so the fabricator could find the defects, often the size of a hair or smaller.

With the establishment of improved data-handling software and the creation of data formats that allow machines to "talk" to each other, the preferred method of generating AOI data is from the basic CAD data supplied by the customer. This method eliminates the need to rely on a golden board and significantly reduces the chances of producing costly incorrect product.

Electrical test data can also now be generated from the CAD reference data. This block of data includes drill data for test fixture ("bed of nails") manufacturing and netlists that identify the individual test circuits. Again, this capability eliminates the need to program the testers to a golden board and ensures a circuit trace has not been inadvertently lost.

FINISHES

In addition to tooling, the designer's manufacturing information includes notes on which finish should be used. Until the mid-1980s, the predominant conductor finish was electroplated and reflowed tin/lead (solder).

As trace widths and spacings decreased, the need to protect against electrical shorts increased. The use of solder mask ("solder resist" is the preferred term) grew to the point that almost all PCBs produced today have solder resist specified. Solder resist is generally an epoxy-based thermosetting resin applied by screen printing. It covers and insulates all the traces and pads, except those to which components will be attached by the customer, and provides protection from the action of solder.

During the mid-1980s, solder mask over bare copper (SMOBC) became the dominant configuration. Most often, the finish required on the pads and exposed traces would be solder, applied by dipping in molten solder then blasting with hot oil or hot air to melt and remove the excess. This process, called solder leveling, is still the method of choice. Some users prefer an organic coating as the surface finish.

BUILDUPS

As a final step, the designer often creates a buildup to give an overall picture of the board he is designing. He makes a sketch of the conductive and insulation materials to be used in the board and the order in which they should be placed ("layed up"). Since varieties of copper thicknesses, epoxy properties, and glass weaves are available, the desired combination must be defined before fabrication begins.

In the case of single- or double-sided boards, the mechanical

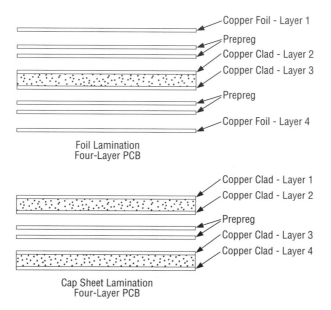

Figure 3-2. Foil lamination versus cap sheet lamination in multilayer boards

drawing provides information that completely defines the buildup. This is not the case for multilayer boards. Defined individually in the buildup, the different layers in multilayer boards will eventually be laminated together by foil lamination or cap sheet lamination. The difference is shown in Figure 3-2.

The agent that bonds the layers of a multilayer PCB together is called prepreg or B stage. Prepreg is glass cloth that has been impregnated with epoxy, then partially cured. It is available in a variety of glass weaves and epoxy compositions. Varying these parameters has a direct effect on the bonding (lamination) process, so selection of proper material is critical.

Sometimes the designer will specifically define the buildup so as to control electrical properties. In other instances, the selection of the prepreg and core is left to the fabricator with only the finished thickness specified.

CHAPTER 4
Manufacturing PCBs

There are 75 to 125 separate operations in the PCB manufacturing sequence, and no two fabricators do things exactly the same way. The differences in processing methods are due to the wide variety of equipment, materials, and proprietary chemistries available to manufacturers. Plus, the size of shops or the individual preferences of shop owners may account for doing some steps differently. This chapter focuses on the predominant manufacturing methods, but some of the common alternative processes will be included in the step-by-step discussions.

The majority of PCBs manufactured in the U.S. use copper-clad epoxy glass as the raw material. The traces (or circuitry) are formed by selectively removing the unwanted copper. This approach is known as subtractive processing.

Additive processing involves the selective deposition of conductive material on a substrate. This approach has been used on a limited basis for several years, but subtractive processing accounts for more than 98% of the PCBs produced in the U.S.

The following sections deal solely with subtractive processing.

PREPRODUCTION

Before manufacturing begins, the fabricator reviews the information and documentation sent by the customer. As discussed in Chapter 3, the customer provides the information either as electronic data (e.g. on magnetic tape or by phone modem) or as films (artworks), mechanical drawings, and specifications. Decisions are then made about the most effective way to manufacture the PCB. These decisions include

- "Number up." The quantity of individual boards arrayed on a particular panel size.
- Panel size to be used. The more common sizes include 18" x 24", 12" x 18", and 20" x 26".
- Features and information to be added during panelization. Items such as UL symbols, test coupons, layer numbers, and borders are selected at this time.
- Material buildup. Often, the individual layer materials are selected by the customer. However, if the composition is not critical, these decisions are left to the fabricator.
- Drilled hole sizes. Since subsequent operations deposit copper in the holes, the holes must be drilled larger than the finished size.
- Tooling holes or target locations. Most fabricators have tooling systems that define the locations of tooling features. This information is often similar for each order and is stored in CAM memory to be used again and again.

The decisions made are summarized by the planning or engineering department on a document called the production traveler (also known as a shop traveler, routing sheet, job order, or production order). The traveler instructs the factory how to process the board. Also included is any other information the factory will need to produce the desired PCB.

DOUBLE-SIDED BOARDS

The following sequence describes the steps used to produce a double-sided, solder mask over bare copper (SMOBC), plated

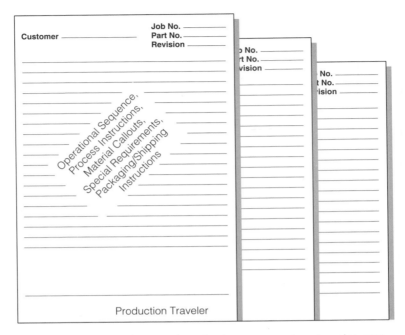

Figure 4-1. Production traveler, also known as a shop traveler, routing sheet, job order, or production order

through-hole (PTH), solder-coated board with gold-plated contacts and component legend.

Although the majority of U.S. output consists of multilayer boards, it is important to first understand the double-sided process. Once a multilayer's internal layers have been fabricated and sandwiched inside the laminate package, a multilayer board is processed almost identically to a double-sided board. Multilayer manufacturing is addressed in Chapter 5.

A single-sided board is also processed like a double-sided board, except that the fabricator starts with raw material clad only on one side with copper and omits Step 5, electroless copper plating.

Note: The illustrations referenced in the following steps appear at the end of the chapter.

Step 1. Material Preparation *(Illustration DS-1)*

Using the information on the traveler, the material necessary to process the order is prepared. The number and size of the panels, the types of materials, and any special instructions are considered at this point.

In the past, the board fabricator bought laminate in large sheets (3' x 4') and then sheared the sheets down to the panel size stated on the traveler. But shearing laminate can produce copper slivers and shreds of epoxy glass that can cause defects at later stages. Today, many fabricators request that laminate suppliers send the laminate precut to panel size. The same is true of the entry material and the backup board.

Step 2. Stack and Pin

The copper-clad panels are stacked (most often three panels to a stack) along with entry material and a backup board. These stacks are then pinned together by drilling two holes (usually ⅛" or ³⁄₁₆" diameter) in opposite ends of the panels and pressing steel dowel pins through the holes. The dowel pins (usually ½" in length) protrude from the stack bottom by approximately ¼" and will eventually locate and hold the stack on the drilling machine table.

Placed on top of the panel stack, the entry material serves several purposes. First, it gives the drill bit a soft material to enter before hitting the copper-clad panel. This improves accuracy and prevents burrs (little spurs of copper created where the drill enters the panel). As the drill retracts from the stack, the entry material helps to clear debris from the drill bit.

Entry material is made from phenolic, aluminum foil, or paper and ranges in thickness from 0.005" to 0.010".

In normal production, holes can be drilled in copper-clad epoxy glass to a depth of five to six times the diameter of the drill without fear of drill breakage or rough holes. The ratio of thickness (or depth) to the smallest hole diameter in a board is called aspect ratio. For example, a drill with a diameter of 0.032" can safely drill a stack depth of 0.160" to 0.192". Using 0.062"-thick material, the measurement equals a stack of three panels.

The backup board, placed on the bottom of the stack, is composed of phenolic, paper composite, or aluminum foil-clad fiber composite. Its purpose is to prevent burrs from forming on the bottom panel and to protect the drill table. Backup is most often 0.093" thick.

Step 3. Drilling (Illustration DS-2)

The stacks (entry material, panels, backup material) are positioned to the drilling machine using the dowel pins previously installed. The dowel pins mate with bushings in the table of the drilling machine so that the stacks are dimensionally located to the drilling machine spindles.

The spindle is the device that holds and turns the drill bit. Today's machines normally have multiple spindles (four or five).

To calculate throughput, or the number of holes a drilling machine can produce per hour, use the following formula:

$$\text{Holes/minute} = \text{Cycles/minute} \times \begin{array}{c} \text{Number} \\ \text{of spindles} \end{array} \times \begin{array}{c} \text{Number of} \\ \text{panels/stack} \end{array}$$

For instance, a five-spindle drilling machine running at 200 cycles (hits) per minute and drilling stacks of three panels can produce:

$$
\begin{aligned}
\text{Holes/minute} &= 200 \text{ cycles/minute} \times 5 \text{ spindles} \times 3 \text{ panels/stack} \\
&= 3{,}000 \text{ holes/minute} \\
&= 180{,}000 \text{ holes/hour} \\
&= \text{more than one million holes per eight-hour shift!}
\end{aligned}
$$

Drill data (the X-Y coordinate information that defines hole position, drilled hole size, spindle speed, drill bit feed, and

Computerized, numerically controlled drilling machine with air-bearing spindles and automatic drill bit changer

retraction rates) is input to the drilling machine by punched paper, magnetic tape, or floppy disk, or through a local area computer network, which downloads the data directly to the machines.

Until the mid-1980s, few drilled holes were less than 0.020" in diameter. Since then, advances in spindle designs, drill bit geometries, and machine stabilities have allowed holes 0.010" and less to be drilled in production environments.

The environment in the drilling department has also changed. Modern drilling machine spindles have air bearings (as opposed to mechanical roller or ball bearings) to enable the

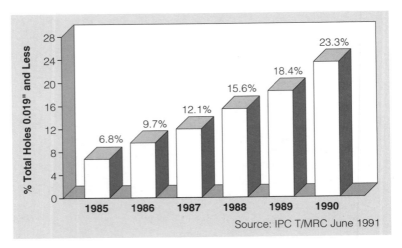

Figure 4-2. Growth of small-hole drilling

spindles to rotate at high speeds—over 100,000 rpm. Air bearings require compressed air that is dry and oil free, often necessitating new or improved air-processing equipment. The computers that run today's drill machines demand spike-free power. The machines themselves, which weigh more than 10,000 pounds, require special floor construction to prevent vibrations from affecting adjacent machines.

Many modern drilling machines are now equipped with automatic load and unload stations and with automatic drill bit changers and broken bit detectors. The automatic load/unload and drill change features allow the machines to run continuously and offer significant labor savings. The broken bit detectors provide assurance that panels with missing holes resulting from a broken bit do not proceed further in the process.

Step 4. Deburr

As the drilling process improves, holes that are virtually burr-free are being produced. Still, most fabricators process drilled panels through a deburring machine of some type. The panels pass through brushes or abrasive wheels that mechanically remove any copper burrs at the rims of the holes.

Some fabricators use optical hole counters immediately after the deburr operation to check for missing, plugged, or extra holes. This operation can also be done manually by using special inspection tools such as panel overlays. An opaque or colored plastic film is drilled with exactly the same pattern as the finished panel. The film is aligned to the drilled panels, and an operator visually checks for missing or extra holes.

Step 5. Electroless Copper Plating (Illustration DS-3)

After drilling, the panels are unstacked, placed on racks, and processed through a series of chemical baths. The chemicals remove any organic contaminants and clean the copper. They also sensitize the epoxy glass on the walls of the drilled holes so it can receive a thin coating of copper. This copper coating alone is not sufficiently thick to carry the electrical load, but it provides a metalized base upon which additional copper can be electrolytically deposited.

Electroless copper depositions range from about 50 microinches to 150 microinches thick. The thicker deposits are called "heavy dep electroless."

Some electroless processes have oxidation-retarding solutions as the final step in the sequence. The decision to apply these solutions depends on the fabricator's preference.

Although electroless copper plating provides the most cost-effective method for metalizing the through holes of a PCB, the process waste is costly to treat. Also, most manufacturers of electroless copper use formaldehyde, a known carcinogen, in the formulation of the chemistry. Several alternative processes were introduced in the 1980s, but none have succeeded in replacing electroless copper as the predominant process.

Exposure of dry-film coated panel using glass phototooling

Step 6. Imaging *(Illustration DS-4)*

In imaging, the circuitry pattern is transferred to the panel. A number of different processes are available to accomplish this step. Regardless of the method used, the result is the same: The PCB panel is covered with a resist, except where copper will ultimately remain on the base material. The fundamental principle of resist is to protect selected areas from further chemical processing, that is, to resist the action of the chemicals.

The three most prevalent processes for applying resist are dry film, screen printed, and liquid.

During the 1960s and early '70s, the resists used were predominantly liquid or screen printed. In the '70s, Du Pont introduced a revolutionary product whereby the resist was manufactured in rolls and could be applied to PCB panels with heat and pressure. During the 1980s and into the '90s, dry film became the most widely used imaging resist.

Dry-film resist. Dry-film resist is an ultraviolet-light-sensitive photopolymer (photoresist). It is supplied on a roll and applied by running the panel through heated rollers (hot roll laminator). After being hot roll laminated, the panel is placed in a UV printer frame, and a phototool, either film or glass, is positioned to the panel using tooling pins as locators. The emulsion on the phototool forms the circuitry pattern plus any auxiliary features added during panelization as discussed in Chapter 3. This emulsion blocks UV light, so the UV light only passes through the clear portions of the phototool to activate the light-sensitive resist. This step is called exposure.

The areas of the dry film exposed to UV light undergo a chemical reaction called polymerization. These areas then become impervious to chemical solutions in the next step of the process—developing.

The panel is placed in a conveyorized developing machine, which sprays developing solutions on the panel to remove the unexposed resist.

Screen-printed resist. Screen printing was an early method of applying resist. These resists are not photosensitive; the pattern is screened directly on the panel.

In this process, a stainless steel mesh screen is coated with a light-sensitive emulsion. The phototool is placed on the emulsion-coated screen, then exposed. The unexposed emulsion is washed away, leaving the circuitry pattern embedded in the screen.

Next, the panel is placed on a screening table, using tooling pins as locators. A squeegee forces resist ink through the open areas of the mesh (where no emulsion is embedded) and onto the panel. The screen-printed panels are then dried in an oven.

Liquid resist. Photosensitive liquid resists are seeing increased use as customers request line widths below 0.005". Liquid resist coatings are much thinner than dry films and by virtue of their physical makeup are able to contour to the irregularities (e.g. pits and dents) in the copper surface of the PCB panel.

One method of applying liquid resists is electrodeposition using the principle of electrophoresis. In this process, the liquid resist and the PCB panel are charged opposite polarities (positive and negative). The resist migrates to the panel and adheres much in the same way as electroplating in Step 7.

After application of the liquid resist, the panels are dried, then exposed and developed in the same manner as dry film.

In the U.S., liquid resists are primarily used in the production of innerlayers. However, their use with double-sided PCBs is increasing, particularly overseas. Innerlayer imaging is discussed in Chapter 5.

Regardless of the imaging process used, the PCB panel is covered with a resist, except where copper will ultimately remain on the base material.

Step 7. Pattern Plating

The panels are clamped in plating racks and immersed in a series of chemical baths that clean the copper pattern (circuitry).

Next, the panels are immersed in copper plating solutions. The solution and panels are electrolytically charged opposite polarities, which causes copper ions to migrate to the uncoated copper areas on the panel. Copper is plated to a thickness of 0.001" on the surface and in the holes. This process takes about one hour. (Illustration DS-5)

The panels are moved from bath to bath either by hand or by

Automatic pattern electroplating line

machine. Automatic plating equipment is computer controlled, and the hoist (or hoists) moves the racked panels through the bath sequence without manual intervention.

The circuitry pattern, now covered with extra copper, is further electroplated with tin or tin/lead (solder). The tin or tin/lead plating sequence is similar to the copper plating process, but only requires five minutes. (Illustration DS-6)

Afterward, the panels are removed from the plating racks. The tin or tin/lead covering the pattern will protect the circuitry in subsequent processing steps.

Step 8. Strip Etch Strip

The panels are now placed in a tank in batches or processed through conveyorized spray equipment to remove the imaging material. This step is called resist stripping. (Illustration DS-7)

After the resist is stripped from the panel, the panel is placed in a conveyorized spray etcher or batch tank. A chemical etchant (usually an ammonia-based compound) attacks—and removes—the uncovered copper but does not attack the tin or tin/lead. The tin or tin/lead resists the actions of the chemical etchant and protects the copper underneath. In this application, the tin or tin/lead is called an etch resist. (Illustration DS-8)

The tin or tin/lead is then chemically stripped from the copper, revealing the circuitry pattern. Since lead has become environmentally undesirable, the use of tin alone increased significantly in the 1980s. (Illustration DS-9)

Step 9. Solder Mask (Illustration DS-10)

A solder mask, more properly termed "solder resist," is a nonconductive coating that is selectively applied to the surface of a PCB to protect certain areas during subsequent operations. The areas left uncoated are pads, lands, and holes to which components are attached, test points so that electrical probes can make contact, and plug-in contact areas as discussed in Step 11.

The most commonly used solder mask materials and their methods of application are

- Epoxy inks applied by screen printing
- Dry-film solder mask laminated to the PCB
- Liquid-photoimageable epoxy inks applied by curtain coating, roller coating, or electrostatic application.

The dry-film solder mask and liquid-photoimageable mask

Liquid-photoimageable solder mask applied by curtain coating

are both processed by exposure to UV light through a film or glass phototool. The unpolymerized material is removed during developing, as in the dry-film imaging process described in Step 6.

Screen-printed solder masks are applied in the same way as the screened primary images, also described in Step 6. The ink is then oven cured or cured using UV light.

Small holes and SMT designs have increased densities that require more accurate placement of the solder mask pattern. The photoimageable processes use glass or film phototooling that is mechanically located to the panel prior to exposure to

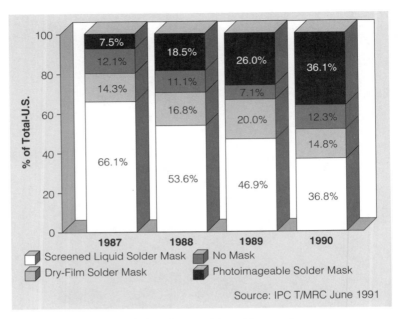

Figure 4-3. Growth of photoimageable solder mask

UV light. Screening, on the other hand, is not as accurate since the mesh of the screen is somewhat elastic and is subject to variances in repeatability. Thus, the use of liquid photoimageable solder resists has grown rapidly in recent years.

Until solder mask over bare copper (SMOBC) became the dominant fabrication method, most boards were made with electroplated solder covering all traces. But when solder mask was applied over a solder-covered surface, it wrinkled and peeled when the solder was heated above its melting point during assembly. SMOBC became the preferred method for this reason.

Step 10. Solder Coating *(Illustration DS-11)*

Ultimately, the PCB user will attach electronic components to the PCB using solder, an alloy of approximately 60% tin and

Horizontal hot-air solder-leveling equipment

40% lead. To facilitate the assembler, the fabricator generally supplies PCBs with solder-coated pads and attachment surfaces, since a solder coat provides the most solderable surface.

Fabricators use several methods of solder coating, but all of them involve dipping the panel into molten solder. The solder coats the pads and holes not covered by solder mask. The excess solder is removed with a blast of hot oil or hot air. However, the hot oil or hot air does not remove the solder that has formed a chemical (intermetallic) bond with the copper.

In either case, the removal of the excess solder is called "solder leveling." The hot-oil process is known as "hydro squeegee." The hot-air process is known as HAL (hot-air leveling) and can be done in vertical or horizontal equipment. Final solder coating thicknesses of 50 to 1,200 microinches can be achieved with most solder-leveling processes.

Step 11. Gold Plating

Some PCBs are designed with contact areas that ultimately mate with a connector or other interconnecting device. Usually the designers call for these contact areas to be plated with nickel and gold. However, some plug-in contact designs do not specify nickel and gold plating and are left either as a solder-coated or bare copper surface. Solder and bare copper do not provide reliable contact surfaces over a long period of time and are used only in specific applications.

The first step in the gold plating process is to mask off all parts of the panel except the area to be plated. A special adhesive tape conforms to the circuit trace geometries and prevents plating solutions from damaging the traces. Because of the way the boards are arranged on the panel, it may be necessary to shear the panel at this point.

The solder is stripped (chemically removed) from the contact areas (called fingers or tabs). The copper is cleaned, then electroplated with nickel and gold. Generally, nickel is plated to 100 to 200 microinches thick and gold to 30 to 50 microinches thick. Compared with the gold, the nickel surface is hard and provides resistance to wear during plugging and unplugging of the PCB assembly.

The gold plating process is also known as "tab plating" or "gold tipping."

If the customer specifies that the boards be shipped in panel form (i.e., not individually routed), then it may be necessary to mask the entire panel except the areas to be gold plated. Dry-film resist is used in this situation. The nickel and gold are plated on all contact areas simultaneously. This process is known as "deep tank gold plating" because it requires tanks deep enough to submerge the entire panel. With tab plating, only the tabs are submerged.

Step 12. Component Legend

Also called "nomenclature," the component legend comprises the identification symbols screen printed on the board to aid the assembly operation. Test or field service personnel may use the legend to locate a particular component on an assembled board.

Component legend can be screen printed on the panel or onto individual boards. Epoxy ink, which comes in a variety of colors, is most often specified. The ink is oven cured after screen printing.

Step 13. Fabricate

The term "fabricate" describes the many mechanical operations that bring the PCB to its final dimensions and that create any specified slots, grooves, bevels, or chamfers. Routing, as defined by the rout data or rout tape made in the tooling cycle, is included.

Some fabricators use punch presses and dies, rather than machine routing, to blank the final board configuration, but the use of this process is decreasing due to cost and inflexibility.

Step 14. Electrical Test

Using the CAD-generated test data (described in Chapter 3), a test fixture is made. A test fixture is usually constructed by drilling a piece of nonconductive material (such as acrylic or epoxy glass) with the same pattern as the drilled holes in the

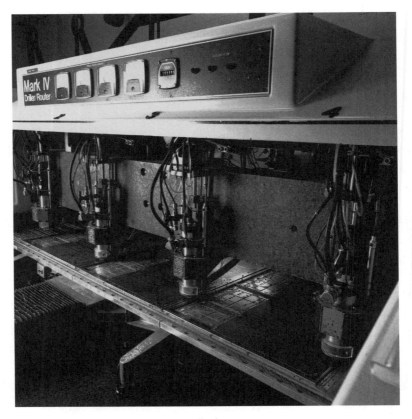

Computerized, numerically controlled router

PCB to be tested. For SMT boards, the pattern of the pads on which the components will be mounted is also drilled.

Spring-loaded metal pins are inserted in the drilled holes of the test fixture to make contact with the corresponding points on the PCB to be tested. The opposite ends of the metal pins are connected by wire to the test equipment. This construction forms a "bed of nails" onto which the PCB to be tested is placed.

The test fixture is therefore unique to the PCB it is designed to test and cannot be used for any other board. Most bed of nails test fixtures contact one side of the PCB. However, hinged "clam-shell" fixtures that have two beds of nails (one on top and one on bottom) have been developed. Also, some test machines can hold a top fixture and a bottom fixture so as to contact both sides of the board.

The fixture is mounted within the electrical test equipment and the board or panel is placed on the pins. The CAD-generated test program causes the test equipment to pass electrical current between specific pairs of pins. Each specific pair makes contact with the starting and ending points of a network on the PCB. If no current passes, an "open circuit" is present; if current passes to an incorrect pin, a "short circuit" has been detected. The test machine then identifies the particular networks with shorts or opens. Repair may be possible, but many customers allow little or no repair of PCBs by the board fabricator. A short or open at this point would then cause the board to be rejected (or scrapped).

Today's programs and equipment can test at low or high voltage and can test insulation resistance. Insulation resistance is a measure of the PCB's ability to withstand voltage applied to adjacent circuitry on the PCB surface without breaking down (arcing). Some of the more sophisticated test equipment can detect small variances in conductor cross-sections. This technique is called "partial opens, partial shorts testing" and is gaining popularity as conductor widths and spaces decrease.

Step 15. Final Inspection

The final step in the manufacturing process is a visual inspection of the finished PCB. Most often, the PCBs are checked for correct mechanical dimensions and are visually examined for overall appearance and other cosmetic parameters. Most PCB fabricators are adopting the operational and quality philosophies discussed in Chapter 6, and as a result, final inspection today is more of a final audit than a full-blown inspection.

VISUAL AIDS: DOUBLE-SIDED SEQUENCE

The following illustrations depict the physical changes a double-sided PCB undergoes during the production cycle. The drawings do not illustrate many of the cleaning, rinsing, drying, or baking operations and do not show any processes after the solder-coating operation (Step 10). The focus is on what happens to the copper foil at the board surface.

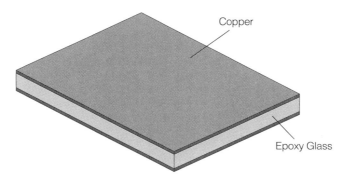

Illustration DS-1. *Copper-clad epoxy glass.* Copper thickness and epoxy type plus panel size are defined on shop traveler (planning).

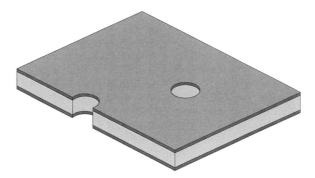

Illustration DS-2. *Drill.* Hole sizes and location are determined by drill data furnished by customer.

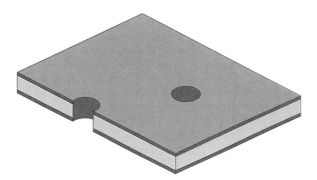

Illustration DS-3. *Electroless copper plate.* A thin layer of copper is deposited on all surfaces including the walls of the drilled holes.

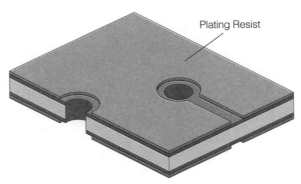

Plating Resist

Illustration DS-4. *Apply plating resist.* The desired circuitry is left uncovered.

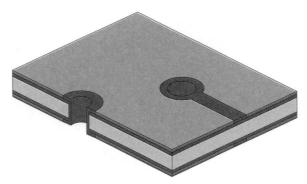

Illustration DS-5. *Electroplate copper.* The specified thickness is electrolytically deposited (usually 0.001").

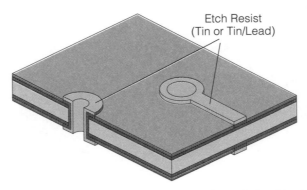

Etch Resist
(Tin or Tin/Lead)

Illustration DS-6. *Electroplate etch resist.* Tin or tin/lead is electrolytically deposited over the copper plating.

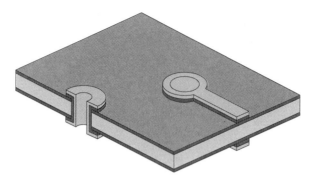

Illustration DS-7. *Strip plating resist*. Plating resist is chemically removed, revealing the surface copper.

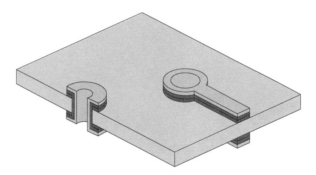

Illustration DS-8. *Etch*. The unwanted copper is removed chemically by an etchant that attacks copper but not tin or tin/lead.

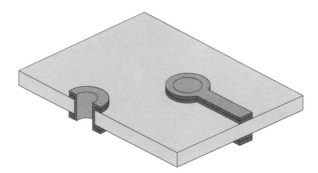

Illustration DS-9. *Strip etch resist*. The tin or tin/lead is chemically removed.

Solder Resist
(Solder Mask)

Illustration DS-10. *Apply solder resist.* The specified resist (dry film, liquid photoimageable, or screen printed) is applied to the surfaces of the PCB or panel.

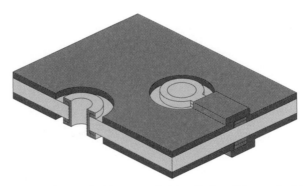

Illustration DS-11. *Solder coat.* Solder (tin/lead) is applied to the exposed copper, and the excess solder is removed.

CHAPTER 5
The Multilayer Process

The largest portion of the U.S. PCB output is multilayer boards (MLBs). Since several layers of circuitry are laminated *within* a multilayer board, this type of board offers more potential interconnections per unit of area for packaging electronic components than do double-sided boards. Up until the mid-1980s, double-sided boards accounted for the bulk of the PCB output; but as the drive for smaller, faster, more powerful electronic products continued, multilayers became the dominant board design.

This chapter describes the steps used to produce a six-layer, plated through-hole (PTH), solder mask over bare copper (SMOBC), solder-coated board with gold-plated contacts and component legend.

To fabricate a six-layer board, two panels of thin, double-sided laminate are imaged and etched to form layers two, three, four, and five. Layers one and six (the outer layers) are formed with copper foil. This technique is called "foil lamination" (Figure 5-1).

In this chapter, the similarities and differences between the steps for a double-sided board and for a multilayer board are discussed when applicable. Once the internal layers have been made and sandwiched inside the laminate package, the pro-

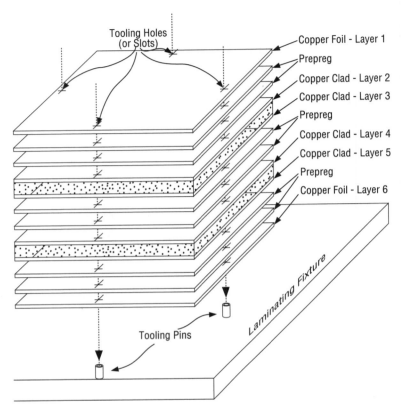

Tooling Holes (or Slots)

Copper Foil - Layer 1
Prepreg
Copper Clad - Layer 2
Copper Clad - Layer 3
Prepreg
Copper Clad - Layer 4
Copper Clad - Layer 5
Prepreg
Copper Foil - Layer 6

Tooling Pins

Laminating Fixture

Figure 5-1. Buildup diagram for a 6-layer multilayer PCB

cessing of an MLB is similar—if not identical—to the processing of a double-sided board.

Note: The illustrations referenced in the following steps appear at the end of the chapter.

MULTILAYER BOARDS

Step 1. Material Preparation (Illustration ML-1)

The thin copper-clad laminate (or core material) used for the innerlayers is sheared (or purchased) to size just like double-sided laminate. The thickness, copper weight, order quantity,

Thin laminate cleaning line

and epoxy-glass designation are defined on the traveler. The innerlayer panels are punched with tooling holes or slots that will subsequently provide reference points for mechanical registration and alignment of the phototooling image.

Step 2. Clean

The innerlayer panels are cleaned to remove any contaminants from the copper surface. The cleaning is done either chemically or mechanically or in some combination of both. Also, reverse current cleaning (de-plating) is gaining popularity. The process is somewhat like electroplating, but the polarities are reversed so that copper ions are drawn from (rather than attracted to) the copper surface.

Step 3. Imaging (Illustration ML-2)

Like double-sided boards, the innerlayers are imaged using dry film, screen printing, or liquid photoresist. However, they are most often printed in reverse: The imaging material covers the copper circuitry desired, not the unwanted copper.

Step 4. Etch Strip (Illustration ML-3)

The innerlayers are then etched in an acid-based or ammoniacal etchant that removes the unwanted copper. The resist is then chemically removed, revealing the copper circuitry.

Step 5. Inspect

Once an innerlayer is laminated inside a multilayer board, an open or short will force the manufacturer to scrap the board, since repair is almost always impossible. As trace widths have decreased in recent years, the visual detection of shorts or opens on innerlayers has become more difficult, increasing the chances of a scrap board after lamination.

Automatic optical inspection (AOI) equipment has gained rapid acceptance since the mid-1980s. It is capable of finding and identifying flaws that would not otherwise be detected. AOI also has measuring capabilities, which make it an ideal medium for statistical quality and process control.

Electronic data representing the circuitry can be generated then downloaded to the AOI machines. This process eliminates the inherent dangers in programming to a "golden" board. This board, thought to be good, could itself be defective, particularly if the original artwork used to produce the board were incorrect. Thus, all boards could compare perfectly to the golden board, but the entire lot would be scrap.

Step 6. Surface Treat (Illustration ML-3)

The copper circuitry on an innerlayer must be treated prior to lamination to improve adhesion to the epoxy-glass bonding

Laying up multilayer boards

agents. Improved adhesion also improves structural strength and overall board reliability.

The most common treatment is black or brown oxide plating. Red oxide or double-treated copper can also be used. Double treat is copper foil that has been prepared at the foil or laminate manufacturer with a chemical treatment and metalization to improve adhesion as the oxide platings do.

Step 7. Layup *(Illustration ML-4)*

Figure 5-1 shows the buildup of a six-layer multilayer package. Layers two, three, four, and five are formed as discussed in Steps 1 through 6. Foils that form layers one and six are placed on the top and bottom of the stack as shown. The operation of creating the stack is called layup.

The partially cured epoxy glass (prepreg) is selected based on customer specification and the information on the traveler. Different glass weaves and epoxy resin contents are available and are defined in detail in Mil-P-13949.

The prepreg provides the bonding material between the layers. Since it is only partially cured, heat and pressure cause the prepreg to flow and bond to the surface of the innerlayers during the lamination operation.

Step 8. Lamination (Illustration ML-5)

The panel layups are stacked—often eight to 10 layups high—and transferred into a laminating press. Some fabricators use sheets of aluminum or stainless steel plus stick-resistant plastic between the individual layups to protect the copper foil surface and ensure that the layups do not stick together. The press applies heat and pressure to the stack, so that the thermosetting resin in the prepreg undergoes molecular cross-linking and bonds all layers together.

Following the heating and pressing cycle (which lasts about an hour), the panels are transferred to a cooling station and are clamped under pressure there until they cool.

The lamination process, and presses in particular, changed dramatically during the 1980s. Vacuum lamination, a method in which the stacks are pressed and heated in a vacuum chamber, increased in popularity during this time. This process reduces the laminating pressure and layer-to-layer slippage and movement. The vacuum is achieved by three basic techniques:

- Vacuum bags. The stacks are enclosed in large plastic bags, called turkey bags, and a vacuum hose is attached.

Lamination presses

- Vacuum frames. The stacks are enclosed in vacuum frames built to fit between the platens of a regular press.
- Vacuum presses. The entire press platen assembly is enclosed and a vacuum chamber is formed.

Another technique gaining popularity is the use of large pressure vessels called autoclaves. Here, the stacks are placed inside and an inert gas (such as nitrogen) or air is introduced under sufficient pressure and heat to achieve lamination.

Step 9. Stress Relief

Next, the panels are placed on baking racks and put in an oven.

Since the panels have been subjected to high heat and pressure, internal stresses have been created that, if not relieved, will cause the panel to warp and twist. Oven heating relieves these stresses. The process lasts for several hours; 325°F for four to eight hours is common.

Step 10. Fabricate Tooling Holes and Trim Edges (Illustration ML-5)

If the panels were fabricated with tooling pins in the layup during the lamination cycle, the panel will have a series of holes along its edges.

If the tooling pins were removed before lamination (and the buildup held together with rivets or glue), there will be no holes along the panel's edges.

In the first situation (called pin lamination), the holes are used as a reference to locate the panels to the drilling machine table. The holes are also used as a reference for trimming the panels to the size specified on the traveler.

In the second case (called glue-lam or rivet lamination), where no holes exist on the panel edge, the tooling holes are formed with X-ray techniques. Pads or targets on the innerlayer are located by X-ray, and tooling holes are punched or drilled.

Another method of forming the tooling holes is "bombsight" drilling to internal target patterns. The patterns are exposed by selectively machining away the copper foil to expose a target on the innerlayer.

Once tooling holes are drilled or punched, the edges of the panel are trimmed using routing (manual or automatic), a diamond saw and template, a milling machine, or other mechanical means. The trimming operation is necessary to eliminate the possibility of any prepreg "flash" (outflow) and to smooth out any uneven or rough edges.

Step 11. Drilling *(Illustration ML-6)*

Drilling multilayers is essentially identical to drilling double-sided boards, except that multilayers must have smooth, smear-free holes. As the drill bit passes through layers of copper and epoxy glass, it generates heat, which can cause the epoxy to smear along the sides of the hole. The subsequent copper plating must form a reliable bond with the internal copper circuitry, so the holes must be free of drilling dust, debris, and smear.

Higher-density MLB designs often have smaller holes than double-sided designs. To maintain quality, smaller stack heights of two (two deep) or even one are used. Stacks for double-sided drilling are typically two to four high since the drill bit sizes are generally larger than for MLBs.

A double-sided board can be given a quick, visual check for proper registration of the circuitry to the drilled pattern. But since MLBs have buried circuitry that is not visible, many fabricators use X-ray machines to check that a drilled hole is properly aligned with an internal pad.

Step 12. Deburr

The same method is used to remove drill burrs from MLBs as is used for double-sided boards, but additional care is needed because of the smaller hole sizes. The smaller holes are more prone to retaining drilling debris.

If holes are clogged, plating solution cannot flow through them and deposit copper. Partial or no copper deposition in the hole means little or no electrical connection to the circuitry on the innerlayers.

Step 13. Electroless Copper Plating *(Illustration ML-7)*

An MLB requires several additional process steps to complete the deposition of copper. These extra steps are necessitated by the epoxy smear that may have been created in the drilling process. This smear prevents the electroless copper from bonding to the copper on the innerlayers, thus creating an open circuit or an unreliable plated though hole. The epoxy smear is removed before electroless plating by immersing the panel in a series of chemical solutions followed by immersion in potassium permanganate or concentrated sulfuric acid. Smear can also be removed by using plasma, a dry-chemical method in which the panels are exposed to oxygen and fluorocarbon gases.

After smear removal, the panels are processed in the same way as double-sided products.

Step 14. Image (Illustration ML-8)

Imaging multilayer panels is identical to imaging double-sided panels. A higher level of dimensional control is required because of surface mount, smaller pads, and higher layer counts. The higher densities make cleanrooms, static control, and temperature and humidity control critical in the imaging area.

Step 15. Pattern Plating (Illustrations ML-9 and ML-10)

As layer counts increase and hole sizes decrease, the aspect ratio (the ratio of the panel thickness to the hole diameter) increases dramatically. This higher ratio demands special techniques and chemistries to achieve uniform plating and adequate thickness in the hole. Specially designed agitation systems and solution movement methods increase the solution flow through the holes, thus improving the plating uniformity. Also, chemistries have been developed to enhance the uniformity and quality of the deposition in smaller holes.

Step 16. Strip Etch Strip (Illustrations ML-11, ML-12, and ML-13)

As conductor widths and spacings decrease, the stripping and etching operations are more difficult. Tighter controls are

Develop/etch/strip line for inner-layer panels

required to remove the imaging material and the underlying copper from between the tightly spaced conductors.

Step 17. Inspect

Because of the smaller lines and spaces, many fabricators perform AOI on dense multilayers prior to solder resist application.

Step 18. Solder Mask *(Illustration ML-14)*

Many of today's MLB designs are not suited to conventional screen printing of solder mask inks because of the tighter dimensional tolerances required. As a result, the use of liquid and dry-film photoimageable solder masks is growing (Figure 4-3).

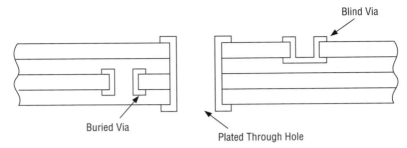

Figure 5-2. Cross sections of blind and buried via holes

Steps 19-24. Solder Coating (Illustration ML-15), Gold Plating, Component Legend, Final Fabrication, Electrical Test, and Final Inspection

Other than the additional care required for a more expensive product, the remaining operations for MLBs have no distinguishing parameters from double-sided boards.

PROCESS AND PRODUCT VARIATIONS

The processes detailed in this chapter represent the methods used to produce the bulk of PCBs in the U.S. today. Each step can be accomplished in a variety of ways with additional variations created by the many different equipment manufacturers and material suppliers.

One significant processing sequence variance comes from a design technique called blind and buried via holes.

- *Blind via.* This plated hole does not extend through the entire thickness of the board, but stops at a given layer.
- *Buried via.* These plated through holes are within innerlayers but do not extend to either external surface.

Manufacturing multilayers that call for buried vias is not particularly difficult. Blind vias, however, require the ability to drill to a given depth of the panel and no deeper (controlled depth drilling). Special drilling equipment is required, and the operation itself is more expensive than regular drilling since the panels cannot be stacked.

With semiconductor packaging demanding more interconnections, the number of designs using blind and buried vias is increasing.

VISUAL AIDS: MULTILAYER SEQUENCE

The following illustrations depict the metamorphosis of the laminate and the surface of the PCB as a multilayer is processed from material preparation through the solder coat operation.

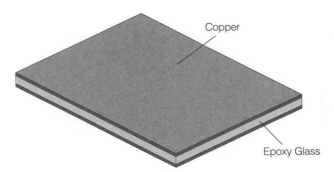

Illustration ML-1. *Copper-clad epoxy glass.* Laminate, core, inner-layer. Copper thickness and epoxy type plus panel size are defined on shop traveler (planning).

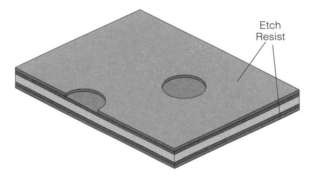

Illustration ML-2. *Apply etch resist, print and develop.* Unwanted copper is now exposed.

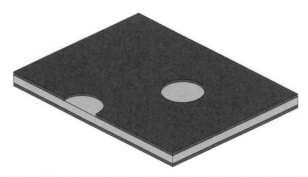

Illustration ML-3. *Etch.* Strip etch resist and treat surface with oxide.

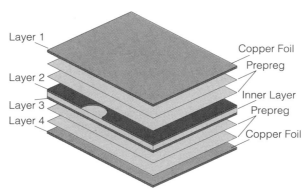

Illustration ML-4. *Layup.* Four-layer multilayer using foil lamination technique. Foil, prepreg type, and size is defined on shop traveler.

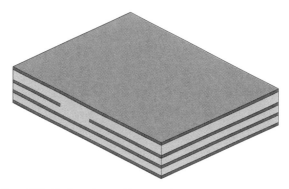

Illustration ML-5. *Laminate.* Heat and pressure cause prepreg to flow and bond layers together.

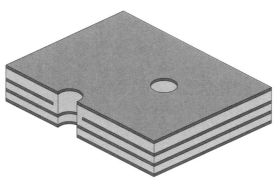

Illustration ML-6. *Drill.* Hole sizes and location are determined by drill data furnished by customer.

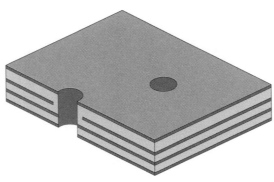

Illustration ML-7. *Electroless copper plate.* A thin layer of copper is deposited following smear removal, cleaning, and preparation.

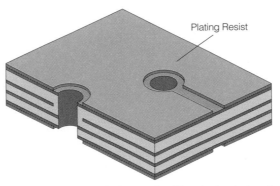

Illustration ML-8. *Apply plating resist.* The desired circuitry is left uncovered.

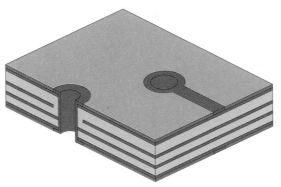

Illustration ML-9. *Electroplate copper.* The specified thickness is electrolytically deposited (usually 0.001").

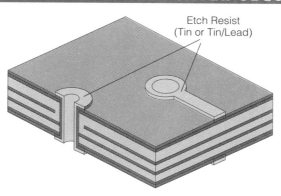

Illustration ML-10. *Electroplate etch resist.* Tin or tin/lead is electrolytically deposited over the copper plating.

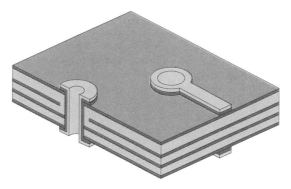

Illustration ML-11. *Strip plating resist.* Plating resist is chemically removed, revealing the copper surface.

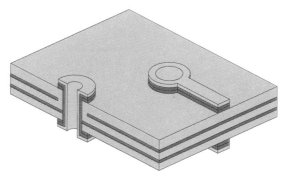

Illustration ML-12. *Etch.* The unwanted copper is removed chemically by an etchant that attacks copper but not tin or tin/lead.

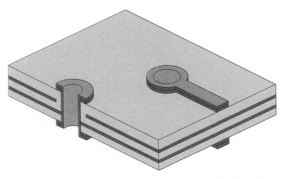

Illustration ML-13. *Strip etch resist.* The tin or tin/lead is chemically removed.

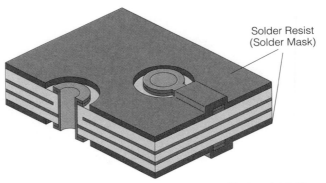

Solder Resist
(Solder Mask)

Illustration ML-14. *Apply solder resist.* The specified resist (either dry film, liquid photoimageable, or screen printed) is applied to the surfaces of the PCB or panel.

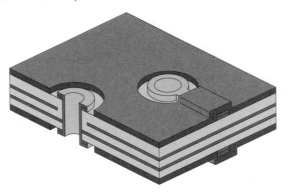

Illustration ML-15. *Solder coat.* Solder (tin/lead) is applied to the exposed copper areas, and the excess solder is removed.

CHAPTER 6

Managing the PCB Business

During the 1980s, offshore competition forced American industry to take a hard look at the way it was doing business. The electronics industry, and specifically the printed circuit industry, realized that the prevailing post-World War II attitudes regarding quality, delivery, and overall customer satisfaction were not working and would not serve well in the future.

The major focus of the redirection in thinking was toward quality. In the past, a certain number of defects were expected, and a certain level of defective product was acceptable. Customers were satisfied if a high percentage of deliveries were made on time. Today's customers demand defect-free products and expect 100% on-time delivery.

These shifts in focus have forced the PCB fabricator to undergo major changes, not only in equipment and processes but also throughout the corporate structure in areas such as staff and human relations, corporate mission, customer satisfaction, supplier relations, and environmental responsibility.

This chapter will deal with some of the specific areas of change and point out significant differences between past and future operating practices in the PCB industry.

QUALITY

Many books, papers, and articles have been published in the last decade on quality and the far-reaching implications of doing business in an environment that expects defect-free products. Procedures and routines have been developed, tested, and instituted—or disregarded. Following are samples of the more significant approaches in the area of quality as they apply to the PCB industry. The combination of these techniques in a corporate culture that focuses on quality and customer satisfaction is referred to as Total Quality Management (TQM).

Statistical Process Control (SPC)

Simply stated, SPC is the use of statistical measuring and analyzing techniques to ascertain that a process is operating within a given set of limits.

Three Sigma

This statistical term relates to a normal distribution curve. Plus or minus three sigma encompasses 99.7% of the data population in a normal distribution curve.

Six Sigma

If zero defective product is the goal, then a statistical tool that encompasses 99.7% of the population (±3 sigma) is not meaningful. Therefore, more PCB producers are establishing routines using ±6 sigma, which covers 99.99% of the population.

PPM

Rather than measure quality levels in percentages (e.g., 96.5%), many fabricators and their customers are measuring parts per million (ppm) and establishing ppm goals. A goal of 1,000 ppm defective would be 99.9% good.

Design of Experiments

Using this technique, a process or product is studied to determine key variables. These variables are then analyzed in depth to reduce the variations they cause. Also, the less important

Quality team meeting

variables are reviewed with an eye toward opening up the tolerances in order to reduce costs.

Often, a PCB fabricator would implement SPC only to discover that the particular process was not under control, even though data gathered indicated otherwise. More and more often, PCB fabricators are using design of experiment techniques as the primary tool in solving chronic quality and yield problems.

Malcolm Baldrige Award

This quality award is the highest bestowed by the U.S. government. Applying for the award, although time consuming, complex, and costly, forces the PCB fabricator to scrutinize the internal workings of the company and identify areas for improvement, upgrade, and change. Some users of PCBs are specifying Malcolm Baldrige application by the fabricator as a condition of doing business in the future.

DELIVERY

As in the quality arena, customer expectations have increased significantly for delivery. Neither late nor early delivery is acceptable. A fabricator's record of on-time shipment is often the basis for qualification.

Just in Time

Just in time (JIT) is an inventory philosophy in which material arrives at the customer site and goes immediately to the factory floor to be used. It reduces the amount of raw or in-process inventory the customer carries, thus reducing costs.

Accomplishing JIT operation requires intense efforts and cooperation between customer and fabricator. A few years ago, early shipment was acceptable and even encouraged, but today many PCB customers specify "zero days late to two days early with no exceptions."

PCB fabricators have adopted similar philosophies as they design shorter production cycles, continuous flow manufacturing, and sole material sourcing into their systems.

Sole sourcing, the concept of having only one supplier of a particular material or service instead of many, has gained popularity, especially for laminates. These partnerships are becoming more common as fabricators and suppliers work together to resolve problems and reduce overall costs.

Dock to Stock

Dock to stock is a delivery technique predicated on the knowledge that the PCB quality is good and can therefore bypass the customer's incoming inspection. Also called "supplier certification," this practice has the obvious advantage of eliminating the incoming inspection step. Achieving such certification requires a strong working relationship between fabricator, customer, and supplier and, although time consuming, has long-term benefits for all.

COST

Once quality and delivery are under control, the PCB fabricator must provide the on-time, defect-free product at a reasonable cost.

Design for Manufacturing

The joint effort between designer and fabricator to minimize the cost of the product is known as design for manufacturing. Some industry experts estimate that up to 25% of the cost of a PCB could be saved if the board were designed for manufacturing. In the past, designers have been less than willing to modify designs to improve yields or overall costs. This hesitancy was precipitated by the fact that manual designs were very time-consuming to modify and changes were prone to errors. Fortunately, with CAD/CAM systems, this trend has disappeared. Now many PCB shops have personnel devoted full time to customer design liaison to reduce costs by improving yields and overall producibility.

Cost Minimization

This system analyzes cost elements and initiates preventive measures to minimize the material, labor, and overhead content of a given design. Before the emergence of multilayers with small holes (less than 0.020"), it was not always necessary to closely scrutinize orders for ways to avoid scrap or improve efficiency and board reliability. However, to meet customer expectations and cost goals, fabricators now conduct producibility or manufacturability reviews before production release. If special testing or experimentation is required to define and quantify an unusual parameter, it is also completed before release.

Waste water treatment equipment

Mandated Costs

Mandated costs incurred by the PCB fabricator are the result of governmental regulation (municipal, state, and federal). Since the mid-1960s, when much of the legislation affecting the PCB industry was passed, fabricators have been required to radically change the way they handle and dispose of hazardous materials and wastes. OSHA (Occupational Safety and Health Act) and EPA (Environmental Protection Agency) regulations place severe penalties on those who do not abide by the laws. It is not unusual for fabricators to spend more than 1 to 2% of revenues for environmental issue compliance.

Although the vast majority of the regulations are intended to enhance the quality of life, they represent an additional cost element to U.S. fabricators that overseas producers may not have to face.

SUMMARY

The PCB industry saw a major shift in the 1980s. Customer expectations rose to demand defect-free product delivered on time at a reasonable cost. Procedurally, fabricators have taken a more scientific approach to quality with the widespread implementation of SPC and design of experiments techniques.

Just-in-time inventory systems and dock-to-stock certification procedures are increasing in popularity as is "partnering" with a customer or supplier.

Covering the costs of environmental issue compliance is an additional challenge. Overseas producers are not bound as tightly by environmental regulations, but this too is changing.

Terms and Definitions*

ADDITIVE PROCESS: A process for obtaining conductive patterns by the selective deposition of conductive material on clad or unclad base material.

ANNULAR RING: That portion of conductive material completely surrounding a hole.

ARRAY: A group of elements or circuits (or circuit boards) arranged in rows and columns on a base material.

ARTWORK: An accurately scaled configuration used to produce the artwork master or production master.

ARTWORK MASTER: The photographic film or glass plate that embodies the image of the PCB pattern, usually on a 1:1 scale.

ASPECT RATIO: A ratio of the PCB thickness to the diameter of the smallest hole.

ASSEMBLY: A number of parts, subassemblies, or any combination thereof joined together.

AUTOMATIC TEST EQUIPMENT: Equipment that automatically analyzes functional or static parameters in order to evaluate performance.

B-STAGE RESIN: A thermosetting resin that is in an intermediate state of cure.

BASE COPPER: The thin copper foil portion of a copper-clad

---

*NOTE: Selected terms and definitions taken from IPC-T-50E

laminate for PCBs. It can be present on one or both sides of the board.

BASE MATERIAL: The insulating material upon which a conductive pattern may be formed. It may be rigid or flexible or both. It may be a dielectric or insulated metal sheet.

BASE MATERIAL THICKNESS: The thickness of the base material excluding metal foil or material deposited on the surface.

BED OF NAILS FIXTURE: A test fixture consisting of a frame and a holder containing a field of spring-loaded pins that make electrical contact with a planar test object (i.e., a PCB).

BLIND VIA: A via hole extending only to one surface of a printed board.

BLISTER: A localized swelling and separation between any of the layers of a laminated base material, or between base material or conductive foil. It is a form of delamination.

BOND STRENGTH: The force per unit area required to separate two adjacent layers of a board by a force perpendicular to the board surface.

BOW: The deviation from flatness of a board characterized by a roughly cylindrical or spherical curvature such that if the board is rectangular, its four corners are in the same plane.

BURIED BIA: A via hole that does not extend to the surface of a printed board.

CHAMFER: A broken corner to eliminate an otherwise sharp edge.

CIRCUIT: The interconnection of a number of devices in one or more closed paths to perform a desired electrical or electronic function.

CIRCUITRY LAYER: A layer of a printed board containing conductors, including ground and voltage planes.

CLAD OR CLADDING: A relatively thin layer or sheet of metal foil that is bonded to a laminate core to form the base material for printed circuits.

Inspecting a cross-section of a multilayer plated through hole

CONDUCTOR: A thin conductive area on a PCB surface or internal layer usually composed of lands (to which component leads are connected) and paths (traces).

CONDUCTOR SPACING: The distance between adjacent edges (not centerline to centerline) of isolated conductive patterns in a conductor layer.

CONDUCTOR THICKNESS: The thickness of the conductor including all metallic coatings.

DEBURRING: Process of removing burrs after PCB drilling.

DEFECT: Any nonconformance to specified requirements by a unit or product.

DEFINITION: The fidelity of reproduction of pattern edges, especially in a printed circuit relative to the original master pattern.

DESIGN RULE: Guidelines that determine automatic conductor routing behavior with respect to specified design parameters.

DESIGN RULE CHECKING: The use of a computer program to perform continuity verification of all conductor routing in accordance with appropriate design rules.

DESMEAR: The removal of friction-melted resin and drilling debris from a hole wall.

DEWETTING: A condition that results when molten solder has coated a surface and then receded, leaving irregularly shaped mounds separated by areas covered with a thin solder film and with the base material not exposed.

DIGITIZING: The converting of feature locations on a flat plane to a digital representation in X-Y coordinates.

DIMENSIONAL STABILITY: A measure of the dimensional change of a material that is caused by factors such as temperature changes, humidity changes, chemical treatment, and stress exposure.

DOUBLE-SIDED BOARD: A printed board with a conductive pattern on both sides.

DRY-FILM RESISTS: Coating material specifically designed for use in the manufacture of printed circuit boards and chemically machined parts. They are suitable for all photomechanical operations and are resistant to various electroplating and etching processes.

DRY-FILM SOLDER MASK: Coating material (dry-film resist) applied to the printed circuit board via a lamination process to protect the board from solder or plating.

ELECTROLESS COPPER: A thin layer of copper deposited on the plastic or metallic surface of a PCB from an autocatalytic plating solution (without the application of electrical current).

ELECTROPLATING: The electrodeposition of an adherent metal coating on a conductive object. The object to be plated is placed in an electrolyte and connected to one terminal of a D.C. voltage source. The metal to be deposited is similarly immersed and connected to the other terminal.

EPOXY: A family of thermosetting resins used in the packaging of semiconductor devices. Epoxies form a chemical bond to many metal surfaces.

ETCHING: The chemical, or chemical and electrolytic, removal of unwanted portions of conductive materials.

HOLE BREAKOUT: A condition in which a hole is partially surrounded by the land.

HOLE PATTERN: The arrangement of all holes in a printed board with respect to a reference point.

INSULATION RESISTANCE: The electrical resistance of an insulating material that is determined under specific conditions between any pair of contacts, conductors, or grounding devices in various combinations.

KGB: Known good board or assembly. Also known as a "golden" board.

LAMINATE THICKNESS: Thickness of the metal-clad base material, single or double sided, prior to any subsequent processing.

LAND: See PAD.

LINE: See CONDUCTOR.

MAJOR DEFECT: A defect that is likely to result in failure of a unit or product by materially reducing its usability for its intended purpose.

MICROSECTIONING: The preparation of a specimen of a material, or materials, that is to be used in metallographic examination. This usually consists of cutting out a cross-section followed by encapsulation, polishing, etching, and staining.

MINOR DEFECT: A defect that is not likely to result in the failure of a unit or product or that does not reduce its usability for its intended purpose.

MULTILAYER PRINTED BOARDS: Printed boards consisting of a number of separate conducting circuit planes separated by insulating materials and bonded together into relatively thin homogeneous constructions with internal and external connections to each level of the circuitry as needed.

PAD: The portion of the conductive pattern on printed circuits designated for the mounting or attachment of components. Also called "land."

PANEL: A rectangular sheet of base material or metal-clad material of predetermined size that is used for the processing of printed boards and, when required, one or more test coupons.

PATTERN: The configuration of conductive and nonconductive materials on a panel or printed board. Also, the circuit configuration on related tools, drawings, and masters.

PATTERN PLATING: The selective plating of a conductive pattern.

PHOTOGRAPHIC IMAGE: An image in a photomask or in an emulsion that is on a film or plate.

PHOTOPLOTTING: A photographic process whereby an image is generated by a controlled light beam that directly exposes a light-sensitive material.

PHOTO PRINT: The process of forming a circuit pattern image by hardening a photosensitive polymeric material by passing light through a photographic film.

PLATED THROUGH HOLE: A hole with plating on its walls that makes an electrical connection between conductive layers, external layers, or both, of a printed board.

Waste treatment plumbing

PREPREG: Sheet material (e.g., glass fabric) impregnated with a resin cured to an intermediate stage (B-stage resin).

PRINTED BOARD: The general term for completely processed printed circuit or printed wiring configurations. It includes single, double-sided, and multilayer boards, both rigid and flexible.

PRINTED CIRCUIT: A conductive pattern that comprises printed components, printed wiring, or a combination thereof, all formed in a predetermined design and intended to be attached to a common base. (In addition, this is a generic term used to describe a printed board produced by any of a number of techniques).

PRINTED WIRING BOARD: A part manufactured from rigid base material upon which completely processed printed wiring has been formed.

REFLOWING: The melting of an electrodeposited tin/lead followed by solidification. The surface has the appearance and physical characteristics of being hot-dipped.

REGISTRATION: The degree of conformity to the position of a pattern, or a portion thereof, a hole, or other feature to its intended position on a product.

RESIN (EPOXY) SMEAR: Resin transferred from the base material onto the surface of the conductive pattern in the wall of a drilled hole.

RESIST: Coating material used to mask or to protect selected areas of a pattern from the action of an etchant, solder, or plating.

SCREEN PRINTING: A process for transferring an image to a surface by forcing suitable media through a stencil screen with a squeegee.

SINGLE-SIDED BOARD: A printed board with conductive pattern on one side only.

SOLDER: An alloy that melts at relatively low temperatures and is used to join or seal metals with higher melting points. A metal alloy with a melting temperature below 427°C (800°F).

SOLDER MASK: Nonpreferred term for solder resist.

STEP-AND-REPEAT: A method by which successive exposures of a single image are made to produce a multiple image production master.

TEST COUPON: A portion of a printed board or of a panel containing printed coupons used to determine the acceptability of such a board.

UNDERWRITERS SYMBOL: A logotype denoting that a product has been recognized (accepted) by Underwriters Laboratories Inc. (UL).

VIA: A plated through hole that is used as an inter-layer connection, but doesn't have component lead or other reinforcing material inserted in it.

VOID: The absence of any substances in a localized area.

WAVE SOLDERING: A process wherein assembled printed boards are brought in contact with a continuously flowing and circulating mass of solder.

Index